CHEMISTRY TERMS

AND

THEIR DEFINITIONS

A POCKET DICTIONARY OF COMMON
TERMS IN CHEMISTRY

KINGSLEY AUGUSTINE

Contents

A 3
B 8
C 10
D 15
E 19
F 22
G 25
H 26
I 29
K 32
L 32
M 33
N 35
O 36
P 38
R 41
S 43
T 47
U 49
V 50
W 50

A

Acid: A substance which releases hydrogen ions when dissolved in water

Acid-base or acid-trioxocarbonate (IV) titrations: They are neutralization reactions in which the colour change of indicators are used to determine the end-point of the titration.

Acid rain: An acidic solution formed when waste gases such as sulphur (IV) oxide dissolve in rain.

Acid salts: Salts formed when the replaceable hydrogen atoms of an acid are only partially replaced by metallic atoms. Examples are $NaHSO_4$ and KH_2PO_4.

Activation energy: The minimum energy needed for a reaction to occur.

Activity series: A list of metals arranged in order of their ability to react with air, water and acids.

Acyclic compounds: Straight chain and branch chain compounds. Examples are butane (straight chain), 2-methylpropane (branched chain).

Addition reaction: A reaction in which an unsaturated compound becomes more saturated by adding on atoms.

Addition polymerization: The polymerization of two or more of the same monomers link together without loss of any molecule.

Adsorption: The ability of some solids to hold gas or liquid molecules on their surface.

Air: A mixture of gases composed mainly of nitrogen, oxygen, carbon (IV) oxide, noble gases and water vapour.

Air pollution: The release of undesired substances known as pollutant into the air thereby causing harm to humans, plants, animals and the environment.

Alkali: A base which dissolves in water.

Alkanes: Saturated aliphatic hydrocarbons which contain only single bonds. Members of the alkane homologous series have the general formula, C_nH_{2n+2}, where n represent the number of carbon atoms.

Alkanoates (or esters): Organic compounds formed by the reaction between alkanols and alkanoic acids.

Alkanoic acids: Organic acids whose functional group is the carboxyl group, -COOH. The general molecular formula of monocarboxylic acids (i.e. having only one –COOH) is $C_nH_{2n+1}COOH$ where n starts from 0, or $C_nH_{2n}O_2$ where n starts from 1.

Alkanols: Organic compounds with one or more hydroxyl (-OH) functional groups linked to a carbon atom. Alkanols with a single hydroxyl group have the general formula $C_nH_{2n+1}OH$ or $C_nH_{2n+2}O$.

Alkenes: Unsaturated hydrocarbons which contain one or more double bonds between the carbon atoms. Members of the alkenes homologous series have the general formula, C_nH_{2n}, where n is 2 and above. The double bond in alkenes acts as its functional group.

Alkynes: Unsaturated hydrocarbons that contain one or more triple bonds between the carbon atoms. Alkynes conform to the general molecular formula C_nH_{2n-2}, where n is a value from 2 and above.

Allotropy: The ability of an element to exist in more than one form in the same physical state.

Alloy: A substance which is a mixture of two or more metals.

Alpha particle: A helium nucleus containing 2 protons and 2 neutrons.

Amides: Organic compounds that are characterized by the presence of the functional group $-CONH_2$ which is called carboxaimde group.

Amines: Organic compounds characterized by the presence of the functional group called amino group, $-NH_2$.

Amphoteric: The ability to show the properties of an acid and a base.

Anhydrous: Does not contain any water of crystallization.

Anion: An ion with a negative charge; anions are attracted to the anode.

Anode: The electrode where electrons leave the electrolyte.

Aqueous solution: A solute dissolved in water.

Aromatic hydrocarbons: Cyclic compounds which are made up of the benzene ring. Examples are benzene and chlorobenzene.

Artificial transmutation: The production of many different atoms by bombarding various elements with fast-moving atomic particles like neutrons, protons, deuterons and alpha particles.

Atmosphere: The gases which surround the earth.

Atmospheric pressure: The pressure exerted by the gases in the atmosphere.

Atom: The smallest particle which shows the properties of an element.

Atomic number: The number of protons in the nucleus of an atom of an element.

Atomic radii: Half the distance between the nuclei of two atoms of the same element in a molecule.

Atomicity: The number of atoms in one molecule of an element or a compound.

Atomization energy: The energy change when one mole of gaseous atoms is formed from the element under standard conditions.

Avogadro's law: It states that equal volume of all gases at the same temperature and pressure contain the same number of molecules.

Avogadro's number/constant: is the number of atoms or molecules or ions present in one mole of a substance. It is equal to 6.02×10^{23} atoms.

B

Base: A substance which will neutralize an acid to form a salt and water only. It is either a metallic oxide or a hydroxide.

Basicity: The basicity of an acid is the number of replaceable hydrogen atoms in one molecule of the acid.

Basic Salts: Salts that contain the hydroxide ion, OH^-. They are formed when an acid does not completely neutralize a base. Examples of basic salts are $Ca(OH)Cl$, $Zn(OH)Cl$ and $Mg(OH)NO_3$.

Beta particle: An electron emitted from the nucleus of an atom.

Biodegradable: Capable of being destroyed by living micro organisms.

Binding energy: The energy evolved in the formation of the nucleus of an element from free protons and neutrons.

Bitumen: Solid residues left behind at temperature above 350^oC during fractional distillation of petroleum. They contain carbon atoms above 70. Bitumen, pitch or asphalt is used for surfacing roads and airfields.

Bomb calorimeter: A device used to determine the heat of combustion of a substance. The word bomb is used to describe it because it is made of strong steel.

Bond energy: The heat energy in kilojoules required to break one mole of covalent bonds.

Boyle's Law: It states that the volume of a fixed mass of gas at constant temperature is inversely proportional to its pressure.

Brine: A concentrated solution of sodium chloride dissolved in water.

Brownian motion: The irregular movements of small particles suspended in a liquid or gas.

Buffer solution: A solution which resists pH changes on the addition of moderate quantities of acid or base.

C

Calibrate: To put a scale on an instrument so that readings can be taken.

Calorific value: The amount of heat produced by the complete combustion of a definite mass of a substance.

Calorimeter: A vessel in which the heat changes during a chemical reaction can be measured.

Carbohydrate: Compounds composed of carbon, hydrogen and oxygen atoms. The hydrogen and oxygen atoms are always in the ratio 2:1, e.g. glucose, sucrose and starch.

Catalyst: A substance which changes the rate of a chemical reaction but is itself unchanged in mass and composition at the end of the reaction.

Catalytic reaction: A reaction which make use of catalyst.

Catalysis: The alteration of reaction rates by catalysts.

Catenation: The formation of covalent bonds between atoms of the same element to form chains or rings.

Cathode: The electrode through which electrons enter the electrolyte.

Cathodic protection: The protection of a metal against corrosion by making it the cathode in an electrolysis cell and coating it with a less reactive metal.

Cation: An ion with a positive charge; cations are attracted to the cathode.

Centrifuge: An apparatus that separates solids from liquids by spinning the mixture at a high speed.

Chain isomerism: Isomers having straight and branched chains.

Charles' Law: It states that the volume of a fixed mass of gas at constant pressure is directly proportional to its temperature in kelvin.

Chemical change: This involves changes which are not easily reversed and in which new substances are formed.

Chemical equilibrium: A position reached in a chemical reaction in which both the forward and backward reactions are talking place at the same time and at the same rate.

Chemical industry: An industry that converts chemical raw materials into useable products.

Chemical reaction: A process in which one or more substances known as reactant undergoes rearrangement of its constituent atoms to form one or more different substances known as product. It is different from a change in physical form or nuclear composition.

Chemistry: The science which deals with the composition, properties, reactions and uses of matter. It is a physical science.

Chlorophyll: The green pigment in a plant that traps light energy for photosynthesis.

Chromatography: Separation of the components of a mixture using a solvent moving on a fixed substance such as paper.

Cloud chamber: A container used to detect radiation by the condensation of a vapour.

Coagulate: To form a solid mass from a solution, to clot.

Coal: A solid formed by the decomposition of dead vegetation over a very long period of time.

Colloid: A mixture having the properties between a solution and a suspension; the solute particles (disperse phase) are intermediate in size between those in a solution and a suspension ($10^{-9} - 10^{-6}$ m).

Combustion: Burning; reaction of a substance usually with oxygen giving out heat.

Common ion effect: The reduction in solubility of a salt in the presence of one of its ions.

Complex ion: An ion which contains a central positive metal ion linked to other atoms, ions or molecules.

Complex Salts: Salts that are formed when two simple compounds are mixed together. Examples are $K_4[Fe(CN)_6]$, $[Cu(NH_3)_4]SO_4$ and Na_2ZnO_2.

Compound: A substance which contains two or more elements chemically combined.

Concentrated Acid: Acid that contains only a little amount of water present in it.

Condensation: Liquid formed from a vapour when cooled.

Condensation polymerization: The polymerization of two or more monomers link together with the loss of a small molecule.

Conductor: A substance which easily allows heat or electricity to pass through it.

Conjugate acid-base pair: Two species related by the loss of a proton, e.g. HNO_3 and NO_3^-.

Coordinate (dative) bond: A covalent bond in which the two electrons have been supplied by the same atom.

Corrosion: The process by which substances in everyday use react with chemicals they come into contact with and are slowly destroyed, e.g. iron rusts when it comes in contact with O_2 and H_2O.

Covalent bond: A force holding atoms together in molecules formed by sharing two electrons between two atoms.

Cracking: The process of breaking down larger hydrocarbons into smaller hydrocarbons.

Crystallization: Formation of crystals when a hot solution cools and becomes saturated at the lower temperatures.

Crystal lattice: The three-dimensional arrangement of the particles of a crystal.

Cyclic compounds: Compounds in which the carbon atoms combine to form a ring. Examples are cyclopropane and cyclohexane.

D

Dative bond or coordinate bond: A covalent bond in which the two electrons have been supplied by the same atom.

Decant: To separate an insoluble solid from a liquid by carefully pouring off the liquid.

Decarboxylation reaction: A type of reaction where the carboxyl group (-COOH) is eliminated.

Decomposition: When a compound breaks up into something simpler.

Dehydrate: To remove water from a substance.

Deliquescence: The process in which water is absorbed from the air to form a saturated solution.

Delocalization: The distribution of negative charge over a number of atoms in a molecule, e.g. in the benzene ring.

Density: Mass per unit volume.

Desiccants: Substances used to dry gases in the laboratory. They are either hygroscopic or deliquescent substances.

Destructive distillation of coal: The heating of coal to a very high temperature in the absence of air so that it does not burn.

Detergent: Synthetic soaps with specific cleaning properties.

Diatomic molecule: A molecule which contains two atoms.

Diffraction pattern: The pattern formed when electromagnetic radiation, such as X-rays are diffracted by passing through a suitable grating.

Diffusion: The process by which gases and liquids will mix completely; the movement of particles of liquid or gas.

Dihydric alkanols or diols: Alkanols with two –OH groups per molecule. An example is ethan-1,2-diol.

Dilute acid: Acid that contains a large amount of water is present in it.

Dipole: A separation of two equal but opposite electrical charges by a small distance.

Dipole-dipole interactions: The attractions between dipoles in different molecules.

Disaccharide: A compound made from two monosaccharide molecules joined together.

Discharge of ions: The loss or gain of electrons at an electrode leading to the formation of products.

Displacement reaction: A reaction in which an atom, ion or group of atoms takes the place of another in a compound.

Disproportionation: The simultaneous oxidation and reduction of an ion or molecule.

Dissolve: When a solute becomes evenly distributed throughout a solvent.

Distillation: A process used to purify liquids or to separate mixtures. The solution is heated and the pure solvent evaporates leaving the solute behind, while the vapour condenses to form the pure solvent.

Divalent: Having a valency of two.

Double bond: Two pairs of electrons shared between two atoms.

Double Salts: Salts that are formed when two simple salts are mixed together. A double salt contains a monovalent metal (e.g. K, Na) and a trivalent metal (e.g. Al and Cr). Examples of double salts are potash alum [$KAl(SO_4)_2.12H_2O$] and chrome alum [$KCr(SO_4)_2.12H_2O$]

Dry cell: An electrochemical cell used to supply energy in which the electrolyte is a paste or similar substance which cannot spill.

Drying agents or desiccants: Substances used to dry gases in the laboratory. They are either hygroscopic or deliquescent substances.

E

Effervescence: Formation of large numbers of gas bubbles in a liquid.

Efflorescence: Loss of water of crystallization from hydrated substances when exposed to the air.

Effusion: Diffusion of gas through a small hole or a capillary tube.

Elasticity: The ability of a substance to change its shape when acted upon by a force and then return to its original shape.

Electrochemical series: A list of elements and ions in order of their standard electrode potentials.

Electrode: A piece of metal or graphite used to carry an electric current into or out of a liquid or solution.

Electrolysis: When an electric current passes through a liquid or solution resulting in chemical changes at the electrodes.

Electrolyte: A liquid or solution which will conduct an electric current by chemical changes at the electrodes.

Electrolytic cell or voltameter: This is a vessel containing two electrodes and an electrolyte.

Electron: A negatively charged particle present in all atoms.

Electron affinity: The energy change when an electron (or mole of electrons) is added to a gaseous atom (or mole of gaseous atoms).

Electron cloud: A region in which electrons move freely.

Electron configuration: The arrangement of electrons in the energy shells of an atom.

Electron shell: An energy level outside the nucleus of an atom where electrons move.

Electronegativity: The ability of an atom to attract electrons to itself when it is in a molecule.

Electrostatic forces: Electrical forces between oppositely charged particles.

Electrovalent bonding: The attractive forces that hold oppositely charged ions together in a lattice.

Element: A substance which cannot be split into simpler substances by chemical means.

Empirical formula: The simplest whole-number ratio of the atoms of the elements present in a compound.

Emulsion: A colloid where both solute (disperse phase) and solvent (disperse medium) are liquids, e.g. milk.

Endothermic reaction: A type of reaction in which heat is absorbed from the surrounding. In this type of reaction, the total heat content of the products is greater than the total heat content of the reactants.

Enthalpy: The enthalpy of a reaction is the heat change which takes place during the reaction.

Entropy: It is a measure of the degree of disorder or randomness of a substance.

Enzyme: A catalyst produced by living cells.

Equilibrium: A reversible reaction is in equilibrium when the rate of the forward reaction is equal to the rate of the backward reaction.

Equilibrium constant: It is a number that shows the relationship between the amount of products and reactants at a particular temperature.

Esterification: The reaction of an organic acid and an alcohol to produce an ester and water.

Evaporation: When particles move faster and leave the liquid phase to form the vapour.

Exothermic reaction: A reaction which gives out heat to the surroundings. It occurs when the total heat content of the products is less than that of the reactants.

F

Faraday: The quantity of electricity (96,500 Coulomb) needed to discharge one mole of atoms from monopositive ions.

Faraday's first law of electrolysis: It states that the mass of an element deposited during electrolysis is directly proportional to the quantity of electricity passed through it.

Faraday's second law of electrolysis: It states that if the same quantity of electricity is passed through different electrolytes, the number of moles of each element deposited is inversely proportional to the charge on the ion of the element.

Fats and oils: The alkanoates of the esterification of propan-1,2,3-triol and fatty acids, which are long-chain alkanoic acids. Fats and oils are also known as tryglycerides or lipids.

Fermentation: The slow conversion of a sugar into ethanol and carbon (IV) oxide, and catalyzed by enzymes.

Fertilizer: A substance added to soil which increases its ability to grow crops by making more nutrients available to the roots of plants.

Filtration: A process used to separate an insoluble solid from a liquid using a filter such as paper.

Fine Chemicals: Chemicals produced in small quantities and to a very high degree of purity. Examples are drugs, perfumes, food additives, dyes, cosmetics, photographic reagents and analytical laboratory reagents such as $AgNO_3$.

First law of thermodynamics: It states that energy can neither be created nor destroyed but can be converted from one form to another.

Floatation: It is a technique used to separate two insoluble solids by mixing them with a liquid such as water, causing one solute to float on the liquid and the other to sink. It can be used to separate a mixture of sawdust and sand.

Fluorescence: The emission of light when exposed to other forms of radiation.

Formula: A way of showing the number of moles of atoms of each element present in a mole of a substance.

Fractional distillation: A process used to separate liquids by evaporation and condensation.

Free energy: The free energy of a system is the energy which is available for doing work.

Freezing: This is the change in state of matter of a substance from liquid to solid.

Frostation: The freezing of a solid-liquid mixture in order to recover the liquid which freezes out.

Functional group: An atom or group of atoms in an organic molecule responsible for the typical properties of the molecule.

Functional group isomerism: Isomers that have different functional group.

G

Galvanize: To cover a metal with zinc to prevent corrosion.

Gamma rays: Electromagnetic radiation produced during radioactive decay.

Gasification of Coke: is the conversion of coke to water gas and producer gas.

Gay Lussac's Law: Gay Lussac's law of combining volume states that when gases react they do so in volumes which are in simple ratio to one another and to the volume of the product formed, if gaseous, provided that temperature and pressure remain constant.

Geiger counter: An instrument used to measure and detect radiation using the ionization of a gas.

Geometric or Cis-trans isomerism: Isomers with two groups put differently on the sides of a double bond.

Graham's law of diffusion: It states that the rate of diffusion of a gas is inversely proportional to the square root of its density provided that temperature and pressure are kept constant.

H

Haemoglobin: The red, iron-containing pigment found in red blood cells which combines reversibly with oxygen to transport the oxygen around the body.

Half-life: The time taken for half of the mass of a radioactive isotope to decay.

Hard water: Water which does not easily form a lather with soap.

Heat capacity: The heat required to raise the temperature of a substance by 1°C.

Heat of Atomization: The heat of atomization of an element is the heat energy absorbed to form one mole of gaseous atoms from the elements. The atomization of an element is an endothermic process since heat is absorbed.

Heat of combustion: The heat given out when one mole of a substance is burnt completely in oxygen. Combustion is an exothermic reaction.

Heat of formation: The heat energy absorbed or evolved when one mole of a substance is formed from its elements.

Heat of neutralization: Heat of neutralization (enthalpy change of neutralization) is the heat give out when an acid reacts with an alkali to form one mole of water. Neutralization is an exothermic process.

Heat of reaction: The amount of heat absorbed or evolved when molar quantities of reactants react to form products.

Heat of solution: The amount of heat absorbed or evolved when a substance is dissolved in excess solvent.

Heavy Chemicals: Chemicals that are produced in very large quantities. Examples of heavy chemical are H_2SO_4, NaOH, NH_3, $CaCO_3$, HNO_3, HCl, Na_2CO_3, bleaching powder and some metals.

Hess' law: It states that the total heat change of a reaction is constant irrespective of the route of the reaction, provided that the reaction conditions are kept constant.

Homologous series: A group of organic compounds which have the same general formula and functional group. They have similar chemical properties and show a gradual change in physical properties with increasing molecular mass.

Hydrated: Contains water of crystallization.

Hydrocarbon: A compound containing only the elements hydrogen and carbon.

Hydrogen bonding: A type of dipole-dipole interaction formed between molecules which have a hydrogen atom covalently bonded to an atom of the very electronegative elements, oxygen, nitrogen or fluorine.

Hydrogenation: Addition of hydrogen to an unsaturated compound.

Hydrogenation of oils: The hardening of oil to produce fat by the addition of hydrogen to oil in the presence of nickel catalyst. Hydrogenation of oil is used for the production of margarine.

Hydrolysis: Reaction of a compound with water.

Hydrolysis of salt: The reaction of a salt with water to form acidic, alkaline or neutral solution.

Hydrophilic: A chemical which dissolves in or mixes well with water.

Hydrophobic: A chemical which does not dissolve in or mix well with water.

Hygroscopic: The ability of substance to absorb water from the air.

Hypothesis: A reasonable explanation proposed for a problem or observation.

I

Immiscible: Liquids that will not mix together.

Indicator: A weak organic acid or base which will produce different colours in a solution according to the hydrogen ion concentration of the solution.

Examples of indicators are methyl orange, methyl red, litmus and phenolphthalein.

Inert-pair effect: The ability of some elements in groups III, IV and V to show oxidation states which are two less than the group number, e.g. tin in group IV shows oxidation state 2.

Insoluble: A substance that does not dissolve in a liquid.

Insulator: A substance which prevents or restricts the passage of heat or electricity.

Intermolecular forces: Attractive forces between molecules

Ion: An atom or group of atoms with an electric charge.

Ion-exchange resin: A solid substance which exchanges some of its ions for those present in a solution, e.g. used to soften water.

Ionic bonding: The attractive forces that hold oppositely charged ions together in a lattice.

Ionic radius: Half the distance between the centres of two ions that are nearest to each other. It is also the distance between the nucleus and the electron in the outermost shell of an ion.

Ionic theory: It proposes that when an electrolyte is melted or it is dissolved in water, it splits into freely-moving positive and negative ions.

Ionization: The process of splitting into ions.

Ionization energy: The ionization energy of an element is the energy needed to remove an electron (or mole of electrons) from a gaseous atom (or mole of gaseous atoms).

Isomers: Compounds having the same molecular formula but different structural formulae.

Isotope: Atoms of the same element having the same numbers of protons but different numbers of neutrons. Or, atoms of the same element having the same atomic number but different mass number.

IUPAC: International Union of Pure and Applied Chemistry.

K

Kinetic theory: A study of the movement of particles in solids, liquids and gases.

L

Latent heat of fusion: The heat that is taken in by a solid as it turns to a liquid.

Latent heat of vaporization: The heat that is taken in by a liquid as it turns to a gas.

Lattice energy: The energy needed to split up a lattice into free gaseous ions separated by an infinite distance from each other.

Law of mass action: It states that at constant temperature, the rate of a reaction is proportional to the active masses of the reactants. Active mass in this law represents concentration.

Ligand: An atom, ion or molecule linked to a positive ion in a complex.

Le Chatellier's principle: It states that if there is a change in temperature, pressure or concentration of a system in equilibrium, the equilibrium will shift in order to neutralize the effect of the change.

Lone pair: Two electrons in the outer shell of an atom which are not involved in bonding.

M

Macromolecule: A molecule made up of many atoms.

Matter: Anything that has mass and occupies space.

Mass: The quantity of matter present in a substance.

Mass number: The sum of the number of protons and the number of neutrons in an atom.

Melting: This is the change in state of matter of a substance from solid to liquid.

Metal: An element whose atom ionizes by electron loss.

Metallic bond: The force of attraction between mobile electrons and the positive nuclei of metal atoms. These bonds hold atoms together in crystal lattice.

Metalloid: An element which shows metallic and non-metallic properties. It is neither a metal nor an insulator but intermediate between metal and insulator

Miscible: Liquids that will completely mix together.

Mixture: Two or more different elements or compounds mixed together but not chemically combined.

Molar volume: The volume occupied by one mole of any gas at a given temperature and pressure. At stp the molar volume is 22,400cm² (or 22.4dm³)

Molarity: The concentration of a solution measured in moles of solute per dm of water.

Mole: A mole of any substance contains 6 x 10^{23} particles.

Molecular formula: The formula which shows the numbers of atoms of each element present in one molecule of the substance.

Molecularity: The molecularity of a reaction is the number of species involved in the rate-determining step of the reaction.

Molecule: A group of two or more atoms joined together by covalent bonds.

Molten: In the liquid state.

Monomer: Small molecules that can join together to form a polymer.

Monosaccharide: A sugar such as glucose which cannot be broken down into simpler sugars.

N

Natural gas: A mixture of gases containing mainly methane which is usually found with crude oil in underground deposits.

Neutralization: The reaction between an acid and a base to form a salt and water.

Neutron: A neutral particle that exists in the nucleus of all atoms except hydrogen.

Non-aqueous solvent: A solvent which is not water or does not contain water, e.g. liquid ammonia.

Normal Salts: Salts formed when all the replaceable hydrogen atoms of an acid are completely replaced by metallic atoms or ammonium ion. Examples are sodium chloride (NaCl) and zinc tetraoxosulphate (VI) ($ZuSO_4$).

Nuclear fission: A nuclear reaction in which a heavy nucleus is split into two atoms of similar size usually accompanied by a large release of energy.

Nuclear fusion: The combination of two nuclei in a nuclear reaction accompanied with a large release of energy.

Nuclear reactor: A complex piece of equipment used to control the energy obtained from nuclear fission so that it can be converted into electricity.

Nucleus: The central part of an atom which contains protons and neutrons.

O

Octane number: Octane number or octane rating is a value used to indicate the ability of a petrol to resist knocking. It is based on a scale on which isooctane (2,2,4 - trimethylpentane) has a value of 100 (minimal knock) and heptane has a value of 0 (bad knock). The higher the octane number of a petrol, the better the quality of the petrol.

Optical isomerism: Isomers of two compounds of the same type whose arrangements of atoms have a mirror image of each other.

Orbital: An energy sub-level which makes up part of a main energy level of an atom.

Order of reaction: The sum of the powers that the concentrations of the reactants are raised to in the rate equation for that reaction.

Organic chemistry: It is the study of the structures, properties, compositions, reactions and preparation of carbon-containing compounds. It is simply the study of the chemistry of organic compounds.

Organic compounds: A large class of chemical compounds in which one or more atoms of carbon are covalently linked to atoms of other elements, most commonly hydrogen, oxygen and nitrogen.

Organic acids: Naturally occurring compounds that have acidic properties. Examples are acetic acid from vinegar, citric acid from oranges, lactic acid from milk and ascorbic acid from fruits and vegetable.

Oxidation: A process in which a substance gains oxygen, loses hydrogen, loses electrons or increases in oxidation number.

Oxidation number, oxidation state: The charge on an ion in a compound or the charge that an atom in a covalent compound could have if it was ionic. It is a positive or negative number assigned to an atom according to a set of arbitrary rules.

Oxidation-reduction reaction: A reaction that involves the transfer of electrons from one atom or ion to another.

Oxidizing agent: A chemical which brings about oxidation.

Oxonium ion, hydroxonium ion: The ion, H_3O^+, formed when hydrogen ions are added to water.

P

Partial pressure: The pressure a gas in a mixture of gases would exert in a vessel if it were alone in that vessel.

Partition law: The ratio of the concentrations of a solute distributed between two immiscible liquids at constant temperature.

Periodic table: A table of all the elements in order of increasing atomic number such that elements with similar properties are arranged in the same vertical column.

Periodicity: The regular variation in chemical and physical properties of the elements in groups and periods of the periodic table.

Permanent hard water: Water containing calcium and magnesium salt in the form of tetraoxosulphate (VI) and chlorides.

pH meter: An electrical machine which records the pH of a solution.

pH scale: A numerical scale used to measure the acidity or alkalinity of solutions.

Photosynthesis: The reaction in which light energy absorbed by chlorophyll in green plants is used to produce carbohydrates and oxygen from carbon(IV) oxide and water.

Physical change: A change which is easily reversed and in which no new substance is formed.

Polar bond: A covalent bond in which the pair of electrons are attracted more to one atom than the other.

Pollution: The release of harmful materials into the environment.

Polymers: are macromolecules formed by joining together a large number of much smaller molecules (monomers) to form a long chain molecule (polymer).

Polymerization: The process in which small molecules are joined together to form a polymer.

Polysaccharide: A compound made from many monosaccharide molecules joined together.

Porous pot: A material which will allow the passage of gases or liquids.

Primary alkanols: Alkanols where the –OH group is attached to the carbon atom at the end of the carbon chain.

Proteins: Long chain molecules containing nitrogen which form an important part of all living organisms.

Proton: A positively charged particle that exists in the nucleus of all atoms.

R

Radiation: The particles and waves produced from the disintegration of a radioactive isotope.

Radical: A group of atoms which behave like a single particle.

Radioactive decay: The continuous disintegration of a radioactive isotope resulting in emission of radiation.

Radioactivity: The spontaneous disintegration of the nucleus of an atom with the emission of radiation. Temperature and pressure have no effect on it.

Redox reaction: A reaction involving oxidation and reduction.

Reducing agent: A chemical which brings about reduction.

Reducing: Sugars that have reducing properties. All monosaccharides and some disaccharides such as maltose and lactose are reducing sugars.

Reduction: A process in which a substance gains hydrogen, loses oxygen, gains electrons or decreases in oxidation number.

Reforming: This is the rearrangement of hydrocarbon atoms to form branched compounds or cyclo-compounds. It is usually carried out by the use of catalyst such as platinum.

Relative atomic mass: The average mass of one atom of an element compared to one twelfth of the mass of one atom of the isotope ^{12}C which is taken to be 12.000g.

Relative molecular mass: The average mass of one molecule of a substance compared to one twelfth of the mass of one atom of the isotope ^{12}C which is taken to be 12.000g.

Relative vapour density: The mass of a definite volume of a gas divided by the mass of an equal volume of hydrogen measured at the same temperature and pressure.

Resonance: A way of representing the distribution of electrons in a molecule using two or more structures.

Respiration: The process by which animals and plants obtain energy by using oxygen to breakdown food materials to carbon (IV) oxide and water.

Reversible reaction: A reaction which can go in both forward and backward directions. It is a reaction which can be reversed due to the reaction between the products to form the original reactants.

S

Salt: A chemical compound formed by the replacement of all or part of the hydrogen ions in an acid with metallic or ammonium ion.

Saponification: The alkaline hydrolysis of fats and oils with caustic alkali to give propan-1,2,3-triol and a sodium or potassium salt called soap.

Saturated compound: A compound in which the valency of the atoms are fully satisfied by single bonds

Saturated solution: A solution which has dissolved the maximum amount of solute at that temperature. It is a solution which contains as much solute as it can dissolve at that temperature in the presence of undissolved solute particles.

Secondary alkanols: Alkanols in which the –OH group is attached to a carbon atom which is bonded to two other carbon atoms.

Semi-permeable membrane: A thin sheet of a substance which will allow the solvent particles but not the solute particles of a solution to pass through it, e.g. cellophane

Second law of thermodynamics: It states that a spontaneous process occurs only if there is an increase in the entropy of a system and its surrounding.

Sieving: A technique used for separating solid particles of different sizes. Gold and diamond in the mining industries are separated from their sediments by sieving.

Single bond: One covalent bond.

Solar heating: A method used to convert the energy of the sun's rays into heat.

Solubility: A measure of how much solute will dissolve in a solvent at a particular temperature, usually given in moles of solute per dm of solvent. It is the number of moles or grams of solute that will saturate $1dm^3$ of solvent at a particular temperature. It is the amount of solute that will dissolve in a unit volume of a solvent to form a saturated solution at a particular temperature.

Solute: The substance that dissolves in a solvent.

Solution: A homogeneous mixture produced when a substance is dissolved in another substance. Also a clear mixture obtained by dissolving a solid in a liquid.

Solvent: The substance which dissolves a solute. It is also a liquid in which a solid is dissolved.

Spectator ion: An ion which is present but does not take part in a chemical reaction

Standard electrode potential: The potential difference set up between a metal and one-molar solution of its ions at $25°C$ relative to the standard electrode potential of hydrogen taken as zero volt. It is a measure of the tendency of atoms or ions in solution to gain electrons.

Standard hydrogen electrode: The electrode which by definition is given the value 0.00V and against which all other electrode systems are measured.

Standard solution: A solution whose concentration is accurately known.

Storage cell: An electrochemical cell (or battery) which can be recharged.

Strong acid: An acid or base which completely dissociates (ionizes) in aqueous solution. Strong acid releases a large amount of hydrogen ion (H^+) in water. Examples are HCl, H_2SO_4 and HNO_3.

Strong Alkali: An alkali that ionize (produce ions) completely in solutions. They produce many hydroxide ions (OH^-) in solution. Sodium hydroxide and potassium hydroxide are examples of strong alkalis.

Structural formula: The formula which shows the arrangement of the atoms in a molecule of the compound.

Sublimation: The change from solid to gas or from gas to solid without passing through the liquid phase. Examples of substances that sublime are iodine, naphthalene and ammonium Chloride.

Substitution reaction: A reaction in which an atom or group of atoms of a molecule is replaced by a different atom or group of atoms.

Supersaturated solution: A solution in which the solvent contains more solute than is necessary to form a saturated solution at that temperature. It is a solution which contains more solute than it normally can hold at that temperature.

Suspension: A mixture of small, insoluble particles in a liquid or a gas.

Synthesis: Formation of a compound from simpler compounds or elements

T

Temporary hard water: Water that contains calcium hydrogen trioxocarbonate (IV), $Ca(HCO_3)_2$.

Tertiary alkanols: Alkanols in which the –OH group is attached to a carbon atom which is bonded to three other carbon atoms.

Thermal dissociation: This is a reversible reaction in which one compound is heated to produce two or more simpler substances.

Thermodynamics: The study of the relationship between heat and other forms of energy.

Thermoplastics: Plastics that can be softened repeatedly by heat and remoulded.

Thermosets: Plastics that cannot be softened or melted by heat or remoulded once they are formed or set.

Titration: A method that is used to determine the exact concentration of a solution by using a burette to add one solution to a known amount of another until the reaction between them is just complete; one of the solutions must be of known concentration. It is the method employed in volumetric analysis whereby a standard solution is used to react with a solution of unknown concentration.

Tracer: A radioactive isotope whose path is monitored when introduced into a living organism or other apparatus under investigation.

Transition metals: The group of elements which lies between group 2 and 3 of the periodic table. They exhibit characteristic properties such as variable valencies, coloured ions, complex ions and their ability to act as catalysts.

Transition series: The three series of elements in the periodic table (Sc to Zn, Y to Cd and La to Hg) which result from the filling in of d-orbitals.

Trihydric alkanols or triols: Alkanols with three –OH groups per molecule. An example is propan-1,2,3-triol (glycerol)

Triple bond: Three pairs of electrons shared between two atoms.

U

Ultraviolet light: Electromagnetic radiation with wavelengths shorter than visible light.

Universal indicator: A mixture of different indicators designed to give a wide range of colours over the pH range 0 - 14.

Unsaturated compound: A compound which contains double or triple bonds and can undergo addition reactions.

V

Valency: The number of hydrogen atoms which one atom of the element will combine with or will replace. It is the combining power of a substance.

Van der Waals forces: Weak intermolecular forces between covalent molecules. It occurs in iodine, naphthalene, ammonium chloride and noble gases.

Vaporization: This is the change in state of matter from liquid to gas.

Vapour: A substance which is in the gaseous state.

Vapour density: It is the ratio of the density of a gas or vapour to that of hydrogen at the same temperature and pressure.

Viscous: A liquid that does not flow easily; sticky.

Voltameter: This is a vessel containing two electrodes and an electrolyte.

Volumetric analysis: Determination of the concentrations of solutions using titrations.

W

Weak acid: An acid or base which only partially dissociates (ionizes) in aqueous solution. A weak acid is an acid which releases a few amount of hydrogen ions in water. Examples are ethanoic acid (CH_3COOH) and trioxocarbonate (IV) acid (H_2CO_3).

Weak Alkali: An alkali that partially ionize in solution. Examples are Calcium hydroxide and ammonia solution.

In order to see other mathematics, physics and chemistry books written by the author, visit: amazon.com/author/kingzohb2. Also, you can simply go to amazon.com and search for the author's name, Kingsley Augustine, and then the books written by the author will show up.

If you have any enquiries, suggestions or information concerning this book, please contact the author through the email below.

Kingsley Augustine

kingzohb2@yahoo.com

www.ingramcontent.com/pod-product-compliance
Lightning Source LLC
Chambersburg PA
CBHW050244230526
45470CB00005B/2105